I0813134

SNAKES

GREEN TREE PYTHONS

by Julie Murray

Cody Koala
An Imprint of Pop!
popbooksonline.com

Hello! My name is Cody Koala

This book is filled with videos, puzzles, games, and more! Scan the QR codes* while you read, or visit the website below to make this book pop.

popbooksonline.com/gtp

*Scanning QR codes requires a web-enabled smart device with a QR code reader app and a camera.

abdobooks.com

Published by Pop!, a division of ABDO, PO Box 398166, Minneapolis, Minnesota 55439.

Printed in the United States of America, North Mankato, Minnesota.

042025
082025

Cover Photo: Shutterstock Images
Interior Photos: AdobeStock; Shutterstock Images
Editors: Elizabeth Andrews and Tyler Gieseke
Series Designers: Neil Klinepier, Colleen McLaren

Library of Congress Control Number: 2024948399

Publisher's Cataloging-in-Publication Data
Names: Murray, Julie, author.
Title: Green tree pythons / by Julie Murray
Description: Minneapolis, Minnesota : Pop!, 2026 | Series: Snakes | Includes online resources and index
Identifiers: ISBN 9781098247829 (lib. bdg.) | ISBN 9781098248369 (ebook)
Subjects: LCSH: Green tree python--Juvenile literature. | Pythons--Juvenile literature. | Constrictors (Snakes)--Juvenile literature. | Snakes--Behavior--Juvenile literature. | Snakes--Juvenile literature.
Classification: DDC 597.96--dc23

Table of Contents

Chapter 1

Where Do They Live?

Green tree pythons are **reptiles**. They live in Australia, New Guinea, and eastern Indonesia. They are found in warm rainforests.

Green tree pythons like places with thick plants and **moisture**.

Green tree pythons spend most of their time in trees. They are often found curled around branches. A green tree python's head sits right in the middle of its curls.

Where Green Tree Pythons Live

Green Tree Python Range

Pacific Ocean

Indonesia

New Guinea

Australia

N

W

E

S

Green tree pythons live alone and only come together to **mate**. They are not **aggressive**, but they will fight if they feel they are in danger. Their bite is painful! They have 100 hook-shaped teeth.

Green tree pythons can be kept as pets, but they do not like to be handled.

Chapter 2

What Do They Look Like?

Green tree pythons are bright green in color. They have broken white or yellow stripes down their back. Some have white, yellow, or blue spots.

Learn more here!

Green tree pythons have long, thin bodies that are covered in smooth **scales**.

They can grow up to 7 feet (2.1m) long and weigh up to 5 pounds (2.3kg).

Green tree pythons have diamond-shaped heads with yellow eyes. They have a **prehensile** tail that is good for living in the trees. The tail also helps the snakes catch **prey**.

Green tree pythons do not have eyelids. They sleep with their eyes open!

Chapter 3

How Do They Hunt?

Green tree pythons are **constrictors**. They often let their tail hang down to attract **prey**. Then, they grab the prey with their sharp teeth.

Explore links here!

Green tree pythons wrap their body around their prey and squeeze until it can't breathe. Then, they swallow their prey whole. Green tree pythons eat mice, birds, and lizards.

Green tree pythons can swallow animals much larger than their heads.

Chapter 4

Green Tree Python Babies

Female green tree pythons lay up to 30 eggs at a time. Eggs hatch within 50 days. Babies are red, yellow, or blue at birth. They will get their green color by six months old.

Complete an activity here!

Making Connections

Text-to-Self

Green tree pythons are green. If you were a snake, what color would you want to be? Why?

Text-to-Text

Have you read other books about snakes? How are those snakes similar to green tree pythons? How are they different?

Text-to-World

Green tree pythons blend into their surroundings. Can you think of any other animals that can do this?

Glossary

aggressive – mean and ready to fight.

constrictor – a snake that kills by wrapping its body tightly around its prey.

mate – to come together to have young.

moisture – a small amount of liquid in the air or on a surface.

prehensile – capable of grasping especially by wrapping around.

prey – an animal that is hunted by other animals for food.

reptile – a cold-blooded animal with a skeleton inside its body and dry scales or hard plates on its skin.

scales – small, hard, thin plates that cover reptiles.

Index

Online Resources

popbooksonline.com

Thanks for reading this Cody Koala book!

This book is filled with videos, puzzles, games, and more! Scan the QR codes* while you read, or visit the website below to make this book pop.

popbooksonline.com/gtp

*Scanning QR codes requires a web-enabled smart device with a QR code reader app and a camera.